Equal Fractions Review

If you multiply the numerator and denominator of a fra[c]
number (larger than 0), you will make an equal fractio[n]

Make equal fractions. First figure out what the numerator of the fraction was multiplied by, and then multiply the denominator by the same number.

2 × [3] = 6 so multiply by [3]

1 × [ ] = 5 so multiply by [ ].

4 × [ ] = 8 so multiply by [ ].

$\frac{2^{\times 3}}{5_{\times 3}} = \frac{6}{15}$ $\frac{1^{\times}}{6_{\times}} = \frac{5}{}$ $\frac{4^{\times}}{7_{\times}} = \frac{8}{}$ $\frac{1}{5} = \frac{4}{}$

$\frac{2}{3} = \frac{16}{}$ $\frac{1}{2} = \frac{7}{}$ $\frac{5}{6} = \frac{20}{}$ $\frac{3}{8} = \frac{9}{}$

$\frac{4}{4} = \frac{16}{}$ $\frac{5}{9} = \frac{10}{}$ $\frac{1}{10} = \frac{2}{}$ $\frac{8}{15} = \frac{24}{}$

Make equal fractions by finding the missing numerators.

$\frac{4}{5} = \frac{}{10}$ $\frac{2}{7} = \frac{}{35}$ $\frac{1}{6} = \frac{}{18}$ $\frac{3}{4} = \frac{}{20}$

$\frac{2}{3} = \frac{}{12}$ $\frac{1}{5} = \frac{}{10}$ $\frac{11}{12} = \frac{}{24}$ $\frac{6}{7} = \frac{}{21}$

The problems below are like the problems above if you "think backwards."

$\frac{^{\times 3}}{4_{\times 3}} = \frac{3}{12}$ $\frac{}{8} = \frac{10}{16}$ $\frac{}{5} = \frac{8}{20}$ $\frac{3}{} = \frac{12}{16}$

$\frac{3}{21} = \frac{}{7}$ $\frac{4}{16} = \frac{1}{}$ $\frac{9}{24} = \frac{}{8}$ $\frac{14}{14} = \frac{}{2}$

Simplest Form

In each row below there are five squares with the same part shaded. Write the fraction for the shaded part below each square. The five fractions are equal fractions. The fraction in simplest form is the fraction that uses the smallest numbers. Circle the fraction in simplest form.

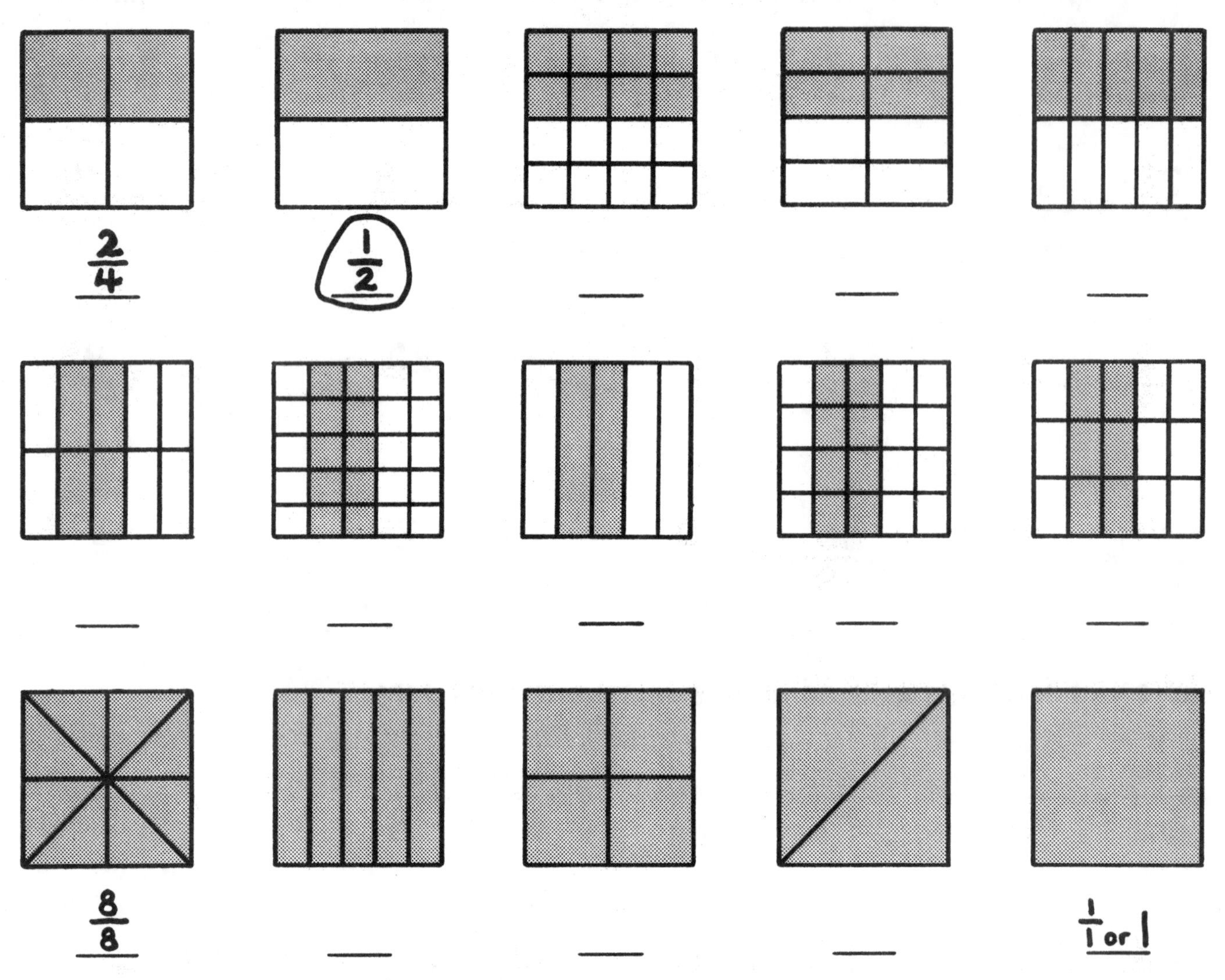

In each group of equal fractions, circle the fraction that is in simplest form.

$\frac{2}{8} = \frac{1}{4} = \frac{6}{24} = \frac{3}{12}$ (with $\frac{1}{4}$ circled)

$\frac{3}{9} = \frac{7}{21} = \frac{1}{3} = \frac{4}{12}$

$\frac{4}{5} = \frac{8}{10} = \frac{12}{15} = \frac{16}{20}$

$\frac{1}{8} = \frac{5}{40}$

$\frac{3}{6} = \frac{1}{2}$

$\frac{5}{7} = \frac{15}{21}$

$\frac{3}{8} = \frac{300}{800}$

$\frac{20}{30} = \frac{2}{3}$

$\frac{3}{21} = \frac{4}{28} = \frac{10}{70} = \frac{1}{7}$

$\frac{1}{10} = \frac{10}{100} = \frac{6}{60} = \frac{3}{30}$

$\frac{16}{44} = \frac{4}{11} = \frac{8}{22} = \frac{12}{33}$

Factors and Products Table

Factors are the numbers multiplied to make a product. Find the missing product of each pair of factors below.

×	1	2	3	4	5	6	7	8	9	10	11	12	13	14	15
1						6	7	8	9	10	11	12	13	14	15
2									18	20	22	24	26	28	30
3														42	45
4													52	56	60
5													65		75
6															90
7											77			98	105
8									72	80	88	96	104	112	120
9													117	126	135
10												120	130	140	
11							77					132	143		
12					60	72	84	96	108	120	132	144			
13						78	91	104	117	130					
14	14	28				84	98	112	126	140					
15	15	30			75	90	105	120	135	150					
16	16	32				96	112	128	144						
17	17	34	51			102	119	136							
18	18	36	54			108	126	144							
19	19	38				114									
20	20					120									
21	21		63		105	126									
22	22		66	88	110	132	154								
23	23			92	115	138	161								
24	24	48				144	168								
25	25	50	75			150	175								
26	26	52	78			156	182								

On the following pages you can use this table in your work with fractions.

Factors

The factors of 28 are all the whole numbers that can be used in multiplication to make 28. You also can think of them as the numbers that divide evenly into 28.

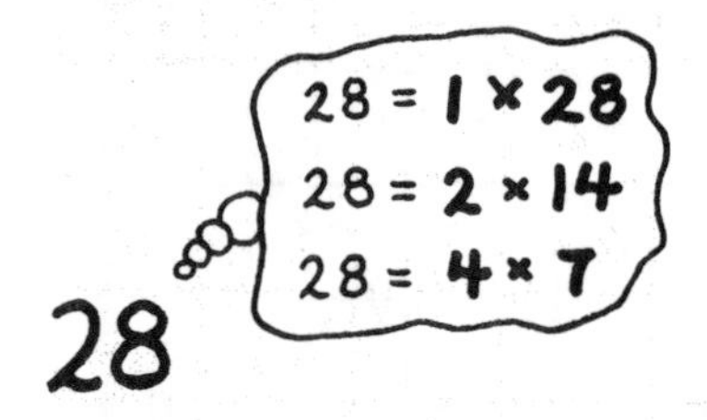

28

The factors of 28 are 1, 28, 2, 14, 4, 7.

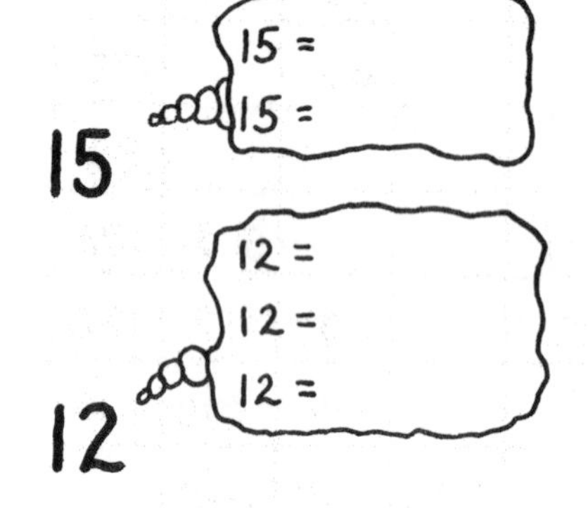

15

The factors of 15 are ___, ___, ___, ___.

12

The factors of 12 are ___, ___, ___, ___, ___, ___.

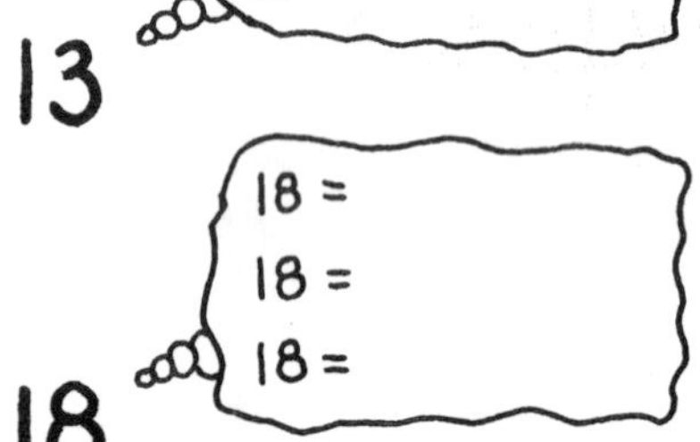

13

The factors of 13 are ___, ___.

18

The factors of 18 are ___, ___, ___, ___, ___, ___.

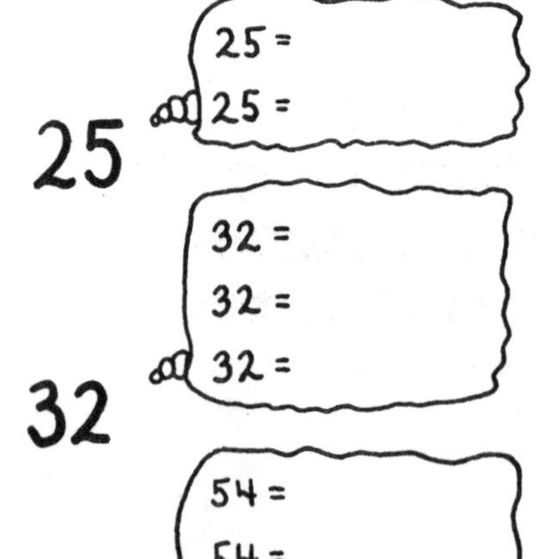

25

The factors of 25 are ___, ___, ___.

32

___, ___, ___, ___, ___, ___ divide evenly into 32.

54 =
54 =
54 =
54 =

54

___, ___, ___, ___, ___, ___, ___, ___ divide evenly into 54.

List all the factors . Use the factors and products table on page 3 if you need help.

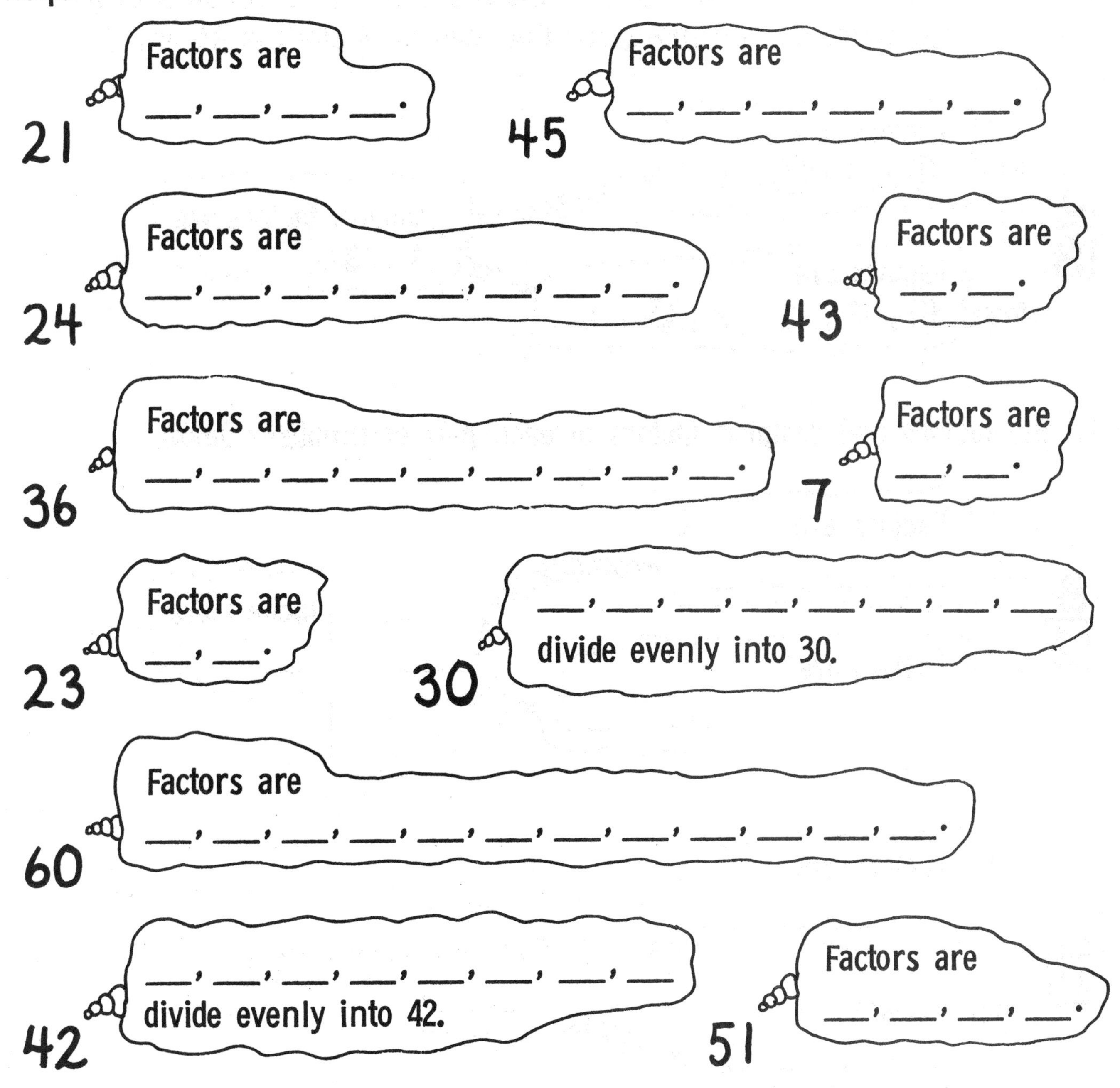

Circle all the factors of the numbers below.

16	1	2	3	4	5	6	7	8	9	10	11	12	13	14	15	16	17	18	19	20
19	1	2	3	4	5	6	7	8	9	10	11	12	13	14	15	16	17	18	19	20
20	1	2	3	4	5	6	7	8	9	10	11	12	13	14	15	16	17	18	19	20

Common Factors

Below are the factors of 15 and 18. 1 and 3 are circled because they are factors of both numbers. 1 and 3 are the common factors of 15 and 18.

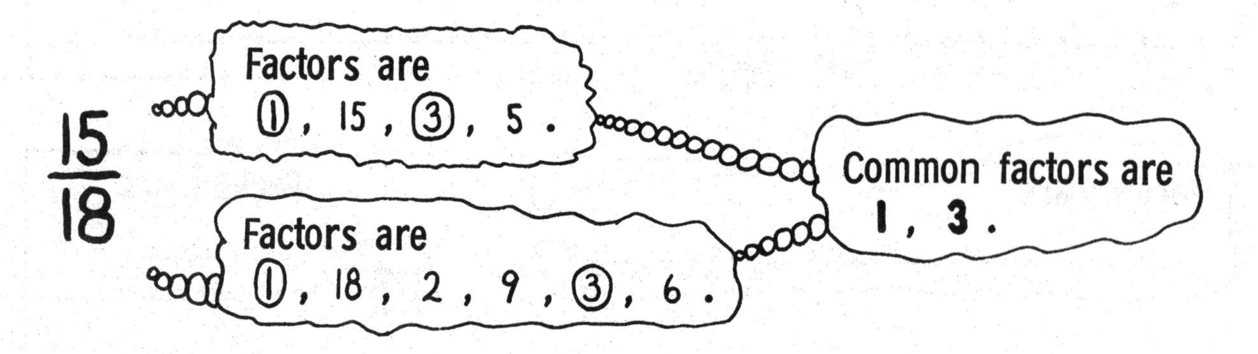

List the factors and common factors of each pair of numbers below.

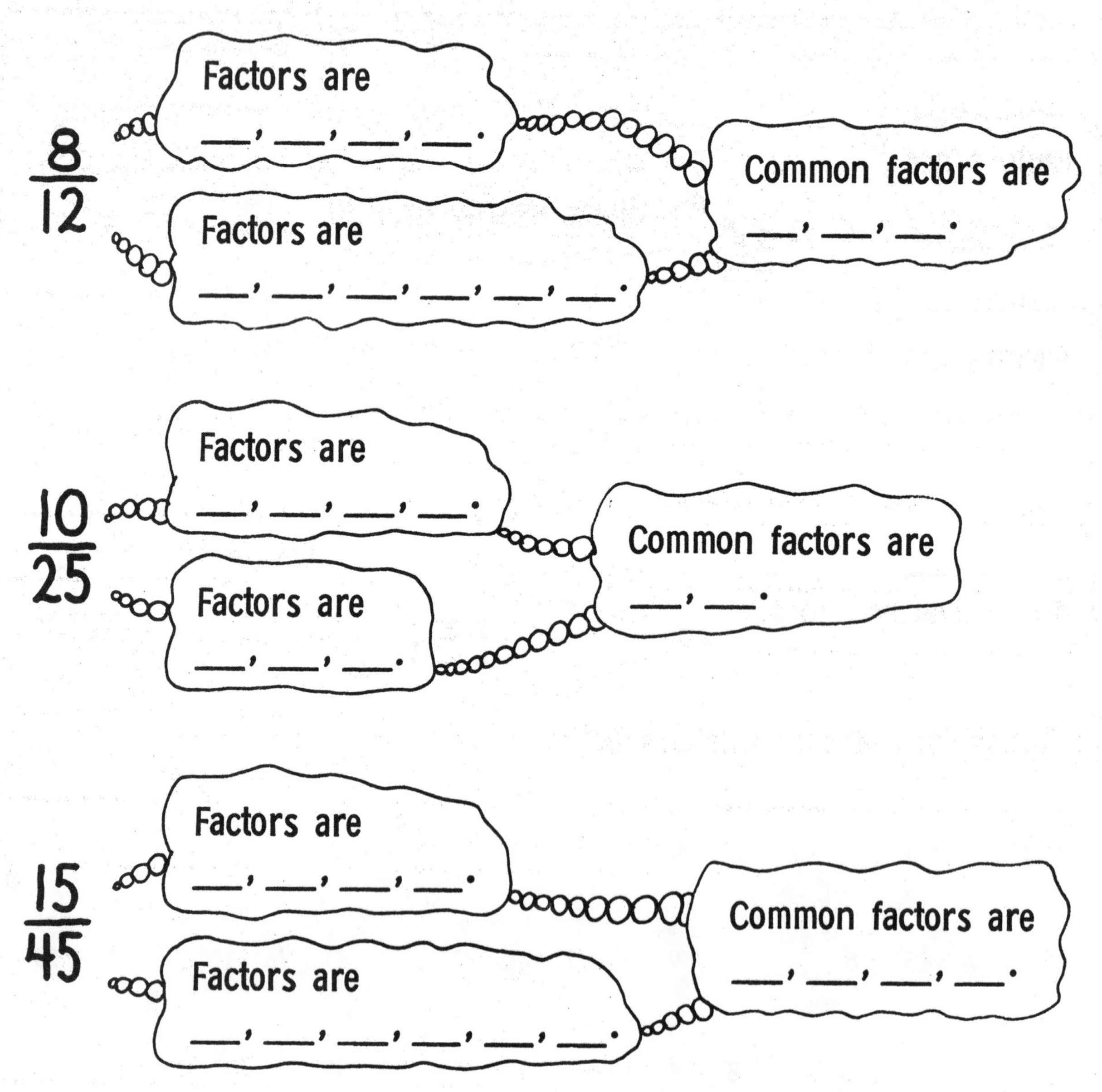

numbers	factors	common factors
12	(1), 12, (2), 6, 3, (4)	1, 2, 4
20	(1), 20, (2), 10, (4), 5	
6	___, ___, ___, ___	___, ___, ___, ___
24	___, ___, ___, ___, ___, ___, ___, ___	
25	___, ___, ___	___, ___
30	___, ___, ___, ___, ___, ___, ___, ___	
27	___, ___, ___, ___	___, ___, ___
36	___, ___, ___, ___, ___, ___, ___, ___, ___	
28	___, ___, ___, ___, ___, ___	___, ___, ___, ___
42	___, ___, ___, ___, ___, ___, ___, ___	
32	___, ___, ___, ___, ___, ___	___, ___, ___ ___, ___,
48	___, ___, ___, ___, ___, ___, ___, ___, ___, ___	

Greatest Common Factor

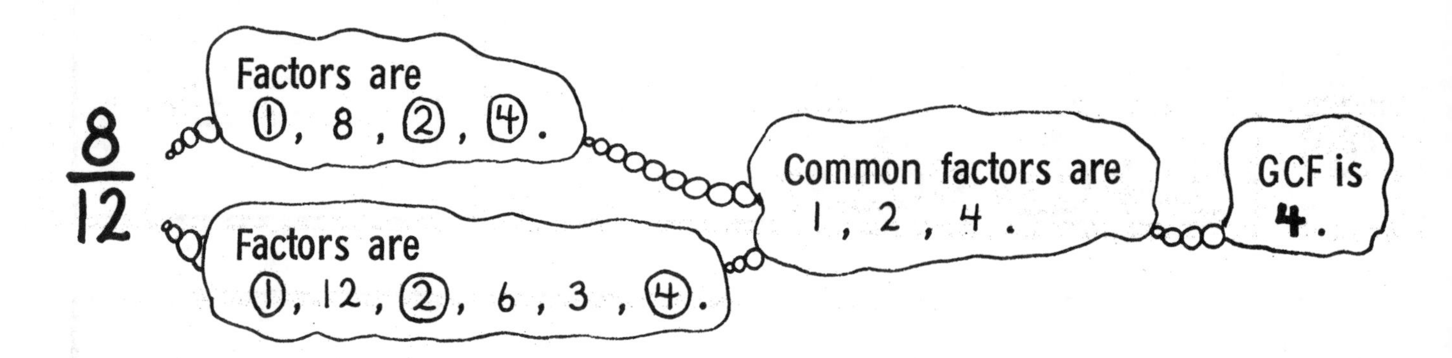

4 is the largest number in the list of common factors. 4 is the greatest common factor (GCF) of 8 and 12. The greatest common factor is the largest whole number that divides evenly into both numbers.

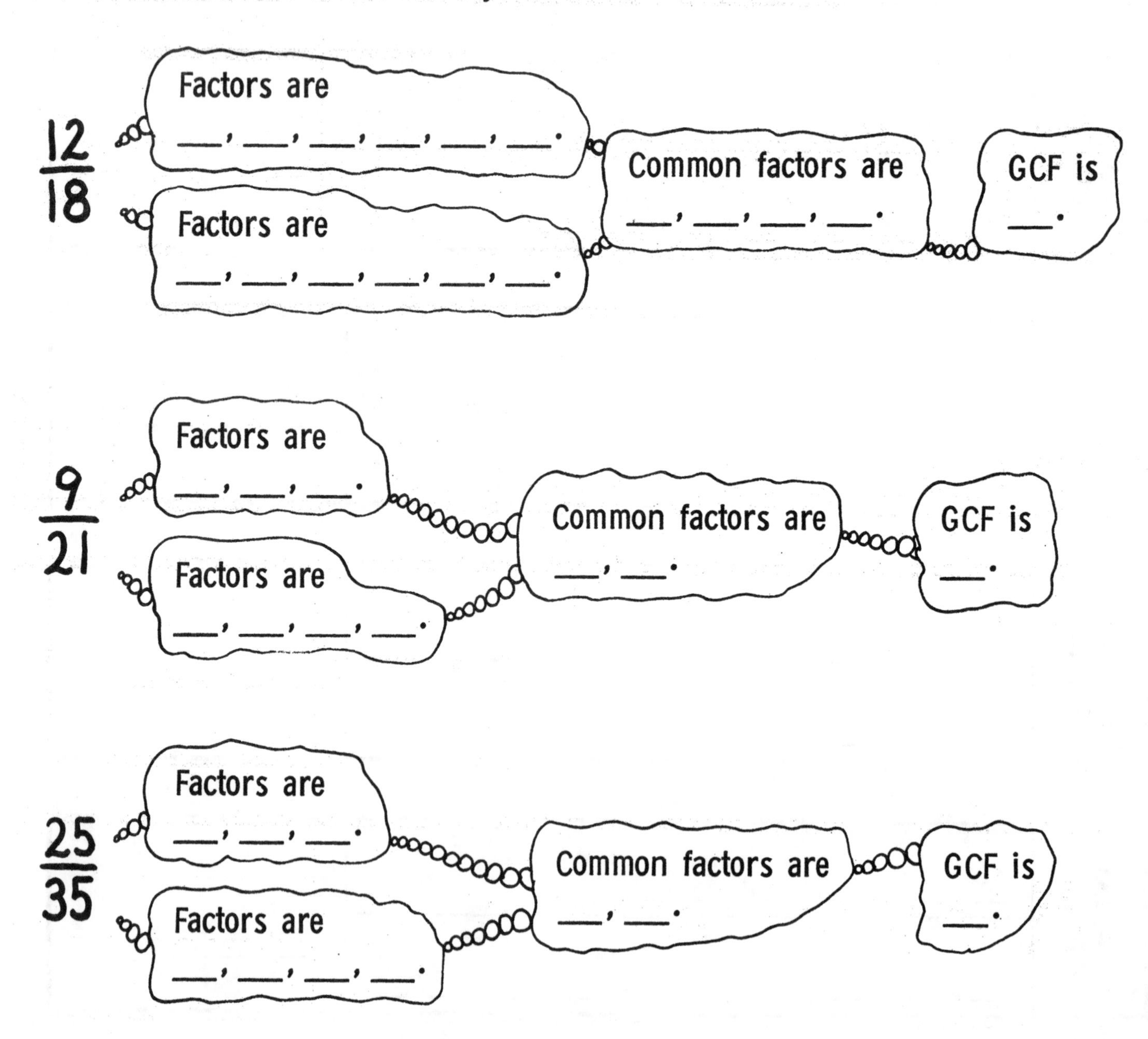

numbers	factors	common factors	GCF
32	___, ___, ___, ___, ___, ___	___	___
45	___, ___, ___, ___, ___, ___		
22	___, ___, ___, ___	___, ___	___
33	___, ___, ___, ___		
40	___, ___, ___, ___, ___, ___, ___, ___	___, ___	___
42	___, ___, ___, ___, ___, ___, ___, ___		
27	___, ___, ___, ___	___, ___, ___	___
36	___, ___, ___, ___, ___, ___, ___, ___, ___		
14	___, ___, ___, ___	___, ___, ___, ___	___
56	___, ___, ___, ___, ___, ___, ___, ___		
60	___, ___, ___, ___, ___, ___, ___, ___, ___, ___, ___, ___	___, ___, ___, ___	___
70	___, ___, ___, ___, ___, ___, ___, ___		

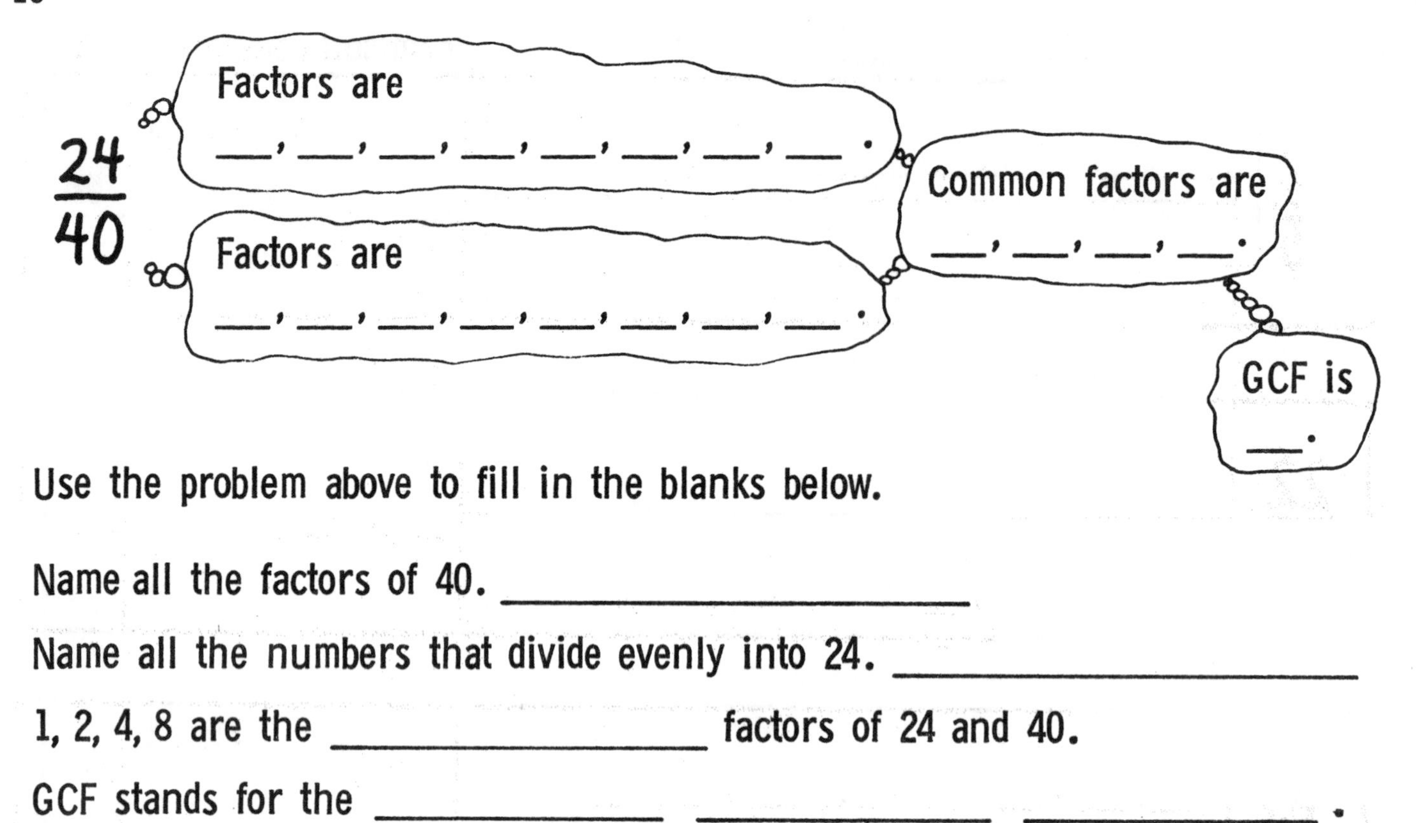

Use the problem above to fill in the blanks below.

Name all the factors of 40. ____________________

Name all the numbers that divide evenly into 24. ____________________

1, 2, 4, 8 are the ________________ factors of 24 and 40.

GCF stands for the ____________ ____________ ____________ .

8 is the ____________ ____________ ____________ of 24 and 40.

Circle all the common factors of each pair of numbers below. The common factors divide evenly into both members of the pair.

$\frac{10}{20}$ ① ② 3 4 ⑤ 6 7 8 9 ⑩

$\frac{16}{24}$ 1 2 3 4 5 6 7 8 9 10 11 12 13 14 15 16

$\frac{18}{27}$ 1 2 3 4 5 6 7 8 9 10 11 12 13 14 15 16 17 18

Look at the common factors above and pick the greatest common factor.

Finding the Greatest Common Factor

Circle all the common factors of the numerator and denominator of each fraction. The common factors are the whole numbers that divide evenly into both the numerator and denominator. Then pick their greatest common factor.

$\frac{12}{30}$ ① ② ③ 4 5 ⑥ 7 8 9 10 11 12 GCF is 6.

$\frac{8}{12}$ 1 2 3 4 5 6 7 8 GCF is ___.

$\frac{15}{30}$ 1 2 3 4 5 6 7 8 9 10 11 12 13 14 15 GCF is ___.

$\frac{6}{18}$ 1 2 3 4 5 6 GCF is ___.

$\frac{7}{8}$ 1 2 3 4 5 6 7 GCF is ___.

Find the greatest common factor of the numerator and denominator of each fraction below. Think of all the whole numbers that divide evenly into both the numerator and denominator and then pick the largest.

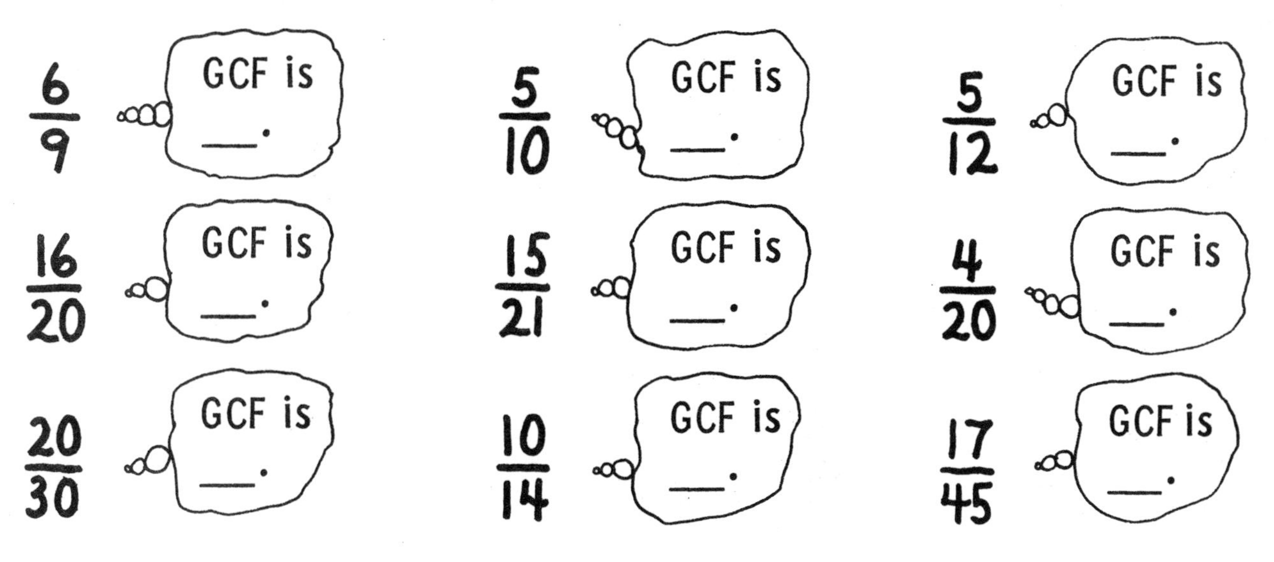

Simplifying Fractions

To rename a fraction in simplest form, you divide both the numerator and denominator by their greatest common factor.

Follow the steps below to simplify $\frac{10}{15}$.

Step 1 $\frac{10}{15}$ (GCF is 5.) Find the greatest common factor (GCF) of the numerator and denominator.

Step 2 $\frac{10 \div 5}{15 \div 5}$ Divide the numerator and denominator by the GCF.

Step 3 $\frac{10 \div 5}{15 \div 5} = \frac{2}{3}$ Rewrite the fraction in simplest form.

Simplify the fractions below.

(GCF is 5.) $\frac{10 \div 5}{15 \div 5} = \frac{2}{3}$

(GCF is 3.) $\frac{3 \div 3}{12 \div 3} =$

(GCF is ____.) $\frac{12}{15} =$

(GCF is ____.) $\frac{14}{21} =$

(GCF is ____.) $\frac{14}{16} =$

(GCF is ____.) $\frac{7}{14} =$

(GCF is ____.) $\frac{5}{35} =$

(GCF is ____.) $\frac{3}{9} =$

(GCF is ____.) $\frac{10}{25} =$

(GCF is ____.) $\frac{2}{20} =$

(GCF is ____.) $\frac{6}{14} =$

(GCF is ____.) $\frac{2}{18} =$

Simplify.

GCF is ___.

$\frac{9}{27} =$

GCF is ___.

$\frac{5}{25} =$

GCF is ___.

$\frac{2}{12} =$

GCF is ___.

$\frac{12}{18} =$

GCF is ___.

$\frac{15}{20} =$

GCF is ___.

$\frac{6}{27} =$

Simplify. Remember to divide the numerator and denominator of each fraction by their greatest common factor.

$\frac{12 \div 3}{21 \div 3} = \mathbf{\frac{4}{7}}$ $\frac{5 \div 5}{20 \div 5} =$ $\frac{4 \div 2}{14 \div 2} =$

$\frac{8 \div 2}{14 \div 2} =$ $\frac{11 \div 11}{22 \div 11} =$ $\frac{15}{25} =$

$\frac{20 \div 10}{50 \div 10} =$ $\frac{14}{35} =$ $\frac{75}{100} =$

$\frac{9}{12} =$ $\frac{12}{20} =$ $\frac{5}{15} =$

$\frac{8}{12} =$ $\frac{4}{20} =$ $\frac{12}{30} =$

Simplify. Try to do the division in your head.

$\frac{10}{16} =$ $\frac{6}{10} =$ $\frac{6}{9} =$

Simplify.

$\frac{10}{22} =$	$\frac{3}{15} =$	$\frac{10}{18} =$	$\frac{3}{6} =$	$\frac{10}{20} =$
$\frac{4}{6} =$	$\frac{4}{24} =$	$\frac{20}{25} =$	$\frac{9}{21} =$	$\frac{4}{40} =$
$\frac{6}{18} =$	$\frac{8}{20} =$	$\frac{12}{18} =$	$\frac{28}{35} =$	$\frac{4}{28} =$
$\frac{12}{14} =$	$\frac{14}{20} =$	$\frac{6}{8} =$	$\frac{5}{10} =$	$\frac{4}{12} =$
$\frac{2}{8} =$	$\frac{6}{21} =$	$\frac{10}{40} =$	$\frac{3}{18} =$	$\frac{2}{22} =$
$\frac{6}{24} =$	$\frac{10}{12} =$	$\frac{4}{10} =$	$\frac{2}{14} =$	$\frac{16}{20} =$
$\frac{20}{22} =$	$\frac{7}{28} =$	$\frac{2}{4} =$	$\frac{7}{21} =$	$\frac{8}{24} =$
$\frac{5}{50} =$	$\frac{6}{16} =$	$\frac{18}{20} =$	$\frac{12}{27} =$	$\frac{8}{16} =$
$\frac{15}{18} =$	$\frac{18}{21} =$	$\frac{8}{10} =$	$\frac{6}{20} =$	$\frac{3}{24} =$
$\frac{30}{35} =$	$\frac{2}{10} =$	$\frac{16}{18} =$	$\frac{6}{15} =$	$\frac{4}{8} =$
$\frac{14}{18} =$	$\frac{10}{14} =$	$\frac{8}{18} =$	$\frac{2}{40} =$	$\frac{8}{22} =$
$\frac{2}{6} =$	$\frac{4}{18} =$	$\frac{3}{21} =$	$\frac{12}{24} =$	$\frac{9}{15} =$

Some fractions can be simplified in one step <u>or</u> in several steps. If you divide by the greatest common factor you will simplify in one step. If you divide by another common factor you will have to use several steps.

$\frac{12 \div 6}{18 \div 6} = \frac{2}{3}$ One step. Divide by the greatest common factor.

$\frac{12 \div 3}{18 \div 3} = \frac{4 \div 2}{6 \div 2} = \frac{2}{3}$ Several steps. Divide by other common factors.

Simplify. Make sure you reach the simplest form for each fraction.

$\frac{9}{18} =$ $\frac{15}{30} =$ $\frac{6}{12} =$

$\frac{12}{16} =$ $\frac{12}{28} =$ $\frac{20}{30} =$

$\frac{16}{24} =$ $\frac{24}{48} =$ $\frac{8}{28} =$

$\frac{4}{16} =$ $\frac{20}{24} =$ $\frac{24}{32} =$

$\frac{18}{24} =$ $\frac{18}{27} =$ $\frac{30}{36} =$

$\frac{15}{40} =$ $\frac{12}{36} =$ $\frac{25}{50} =$

$\frac{8}{32} =$ $\frac{18}{30} =$ $\frac{60}{100} =$

1 and 0 are the simplest names for some fractions.

$\frac{3}{3} = 1 \qquad \frac{4}{4} = 1 \qquad \frac{10}{10} = 1 \qquad \frac{20}{20} = 1 \qquad \frac{44}{44} = 1 \qquad \frac{2000}{2000} = 1$

$\frac{0}{3} = 0 \qquad \frac{0}{4} = 0 \qquad \frac{0}{10} = 0 \qquad \frac{0}{20} = 0 \qquad \frac{0}{44} = 0 \qquad \frac{0}{2000} = 0$

Sandy was asked to simplify $\frac{3}{14}$.

$\frac{3 \div 1}{14 \div 1} = \frac{3}{14}$

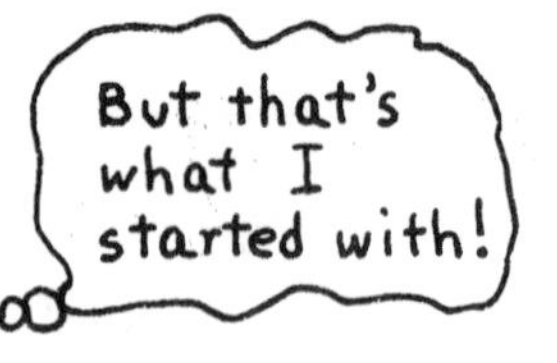

That's right, Sandy. If the greatest common factor of the numerator and denominator is one, then the fraction is already in simplest form. You can't simplify it any more.

Some of the fractions below are already in simplest form. Circle them. Simplify the others.

$\frac{2}{16} = \frac{1}{8}$	$\frac{5}{7}$ (circled)	$\frac{8}{8} = 1$	$\frac{0}{15} = 0$	$\frac{2}{9}$
$\frac{14}{24}$	$\frac{0}{3}$	$\frac{2}{24}$	$\frac{15}{16}$	$\frac{10}{10}$
$\frac{0}{22}$	$\frac{16}{32}$	$\frac{6}{25}$	$\frac{16}{16}$	$\frac{20}{40}$
$\frac{2}{31}$	$\frac{12}{40}$	$\frac{10}{35}$	$\frac{15}{27}$	$\frac{0}{32}$
$\frac{24}{26}$	$\frac{1}{1}$	$\frac{5}{17}$	$\frac{9}{24}$	$\frac{100}{100}$
$\frac{9}{20}$	$\frac{0}{1899}$	$\frac{14}{28}$	$\frac{2361}{2361}$	$\frac{7}{12}$

Chris took the quiz below. Chris was late to school and didn't finish. Put **C** or **X** by each problem that Chris did. Then you do the problems that Chris didn't have time to do.

Chris

Fraction Quiz

Simplify.

$\frac{9}{18} = \frac{1}{2}$ C $\frac{18}{20} = \frac{9}{10}$ $\frac{10}{30}$

$\frac{8}{24} = \frac{2}{6}$ X Not simplest form! $\frac{40}{50} = \frac{20}{25}$ $\frac{21}{35}$

$\frac{9}{24} = \frac{3}{4}$ $\frac{2}{10} = \frac{1}{10}$ $\frac{7}{21}$

$\frac{15}{35} = \frac{5}{7}$ $\frac{24}{30} = \frac{4}{5}$ $\frac{6}{22}$

$\frac{18}{32} = \frac{6}{8}$ $\frac{8}{10} = \frac{4}{5}$ $\frac{25}{40}$

$\frac{8}{16} = \frac{5}{8}$ $\frac{9}{27} = \frac{3}{3}$ $\frac{9}{36}$

$\frac{12}{14} = \frac{6}{7}$ $\frac{150}{200}$

Chris did ____ problems. ____ were correct.
What fraction of the problems that Chris did were correct? ____
What fraction of all the problems did Chris do correctly? ____

Simplifying Fractions with Large Numbers

Fractions with large numbers are hard to simplify because it is hard to find a common factor of the numerator and denominator. Below are some hints to help you simplify fractions with large numbers. You will have to simplify the fractions on this page in several steps. Use the hints to start, then simplify on your own until you reach the simplest form.

$\frac{150 \div 10}{210 \div 10} =$

$\frac{1600}{2000} =$

Hint 1 Divide by 10.

If the numerator and denominator both end in 0, then 10 is a common factor. Divide by 10 as often as you can, then see if the fraction is in simplest form.

$\frac{128 \div 2}{144 \div 2} =$

$\frac{160}{192} =$

Hint 2 Divide by 2.

If the numerator and denominator are both even (end in 0, 2, 4, 6, or 8), then 2 is a common factor. Keep dividing by 2 until they are no longer both even, then see if the fraction is in simplest form.

$\frac{90 \div 5}{105 \div 5} =$

$\frac{125}{625} =$

Hint 3 Divide by 5.

If the numerator and denominator end in 0 or 5, then 5 is a common factor. Divide by 5 as often as you can, then see if the fraction is in simplest form.

$\frac{420}{980} =$

$\frac{850}{1500} =$

$\frac{5600}{8800} =$

$\frac{900}{1350} =$

Use all three hints on these.

Do the numerator and denominator both end in 0? If yes, divide by ___.

Are the numerator and denominator both even? If yes, divide by ___.

Do the numerator and denominator end in 0 or 5? If yes, divide by ___.

Multiplying Fractions

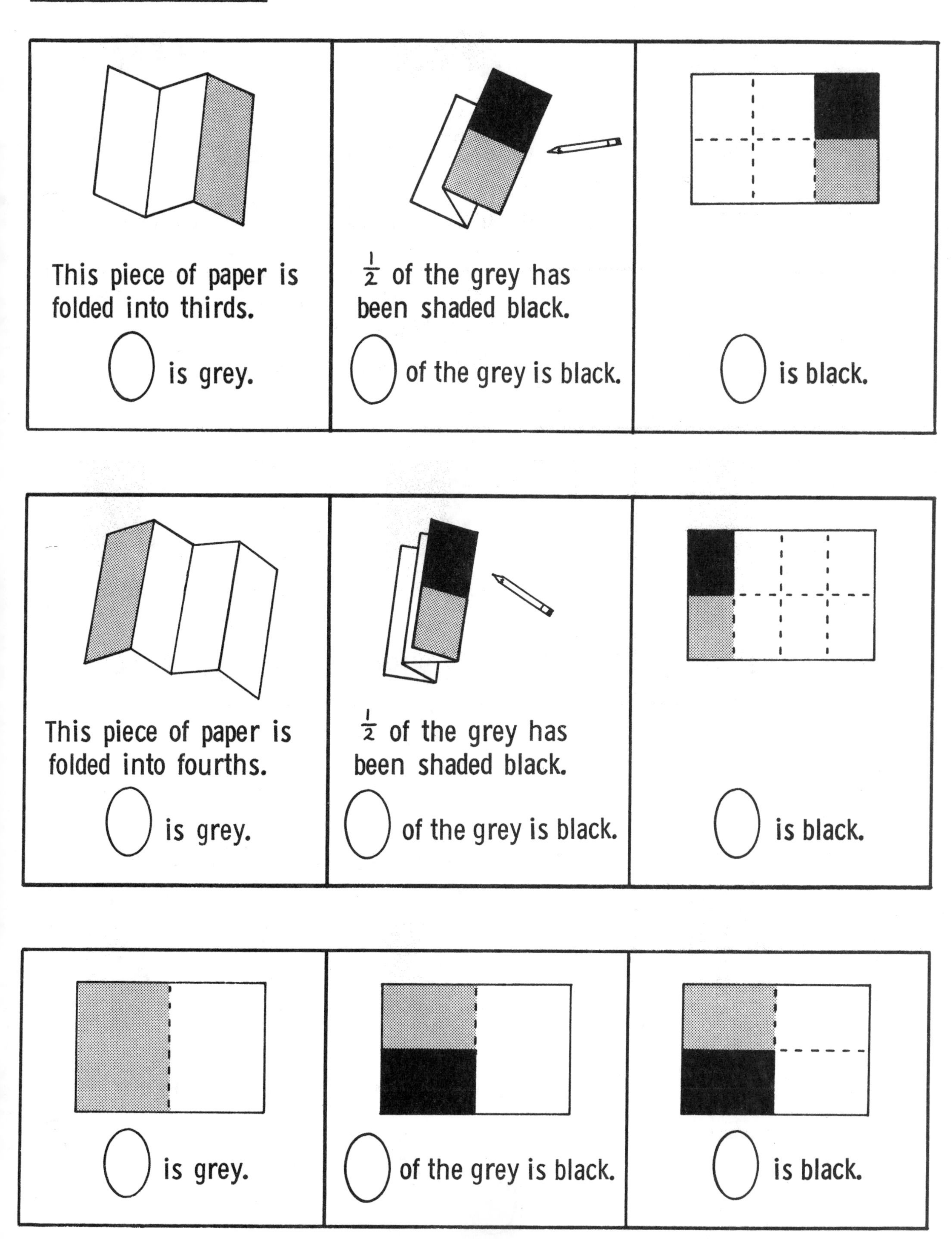

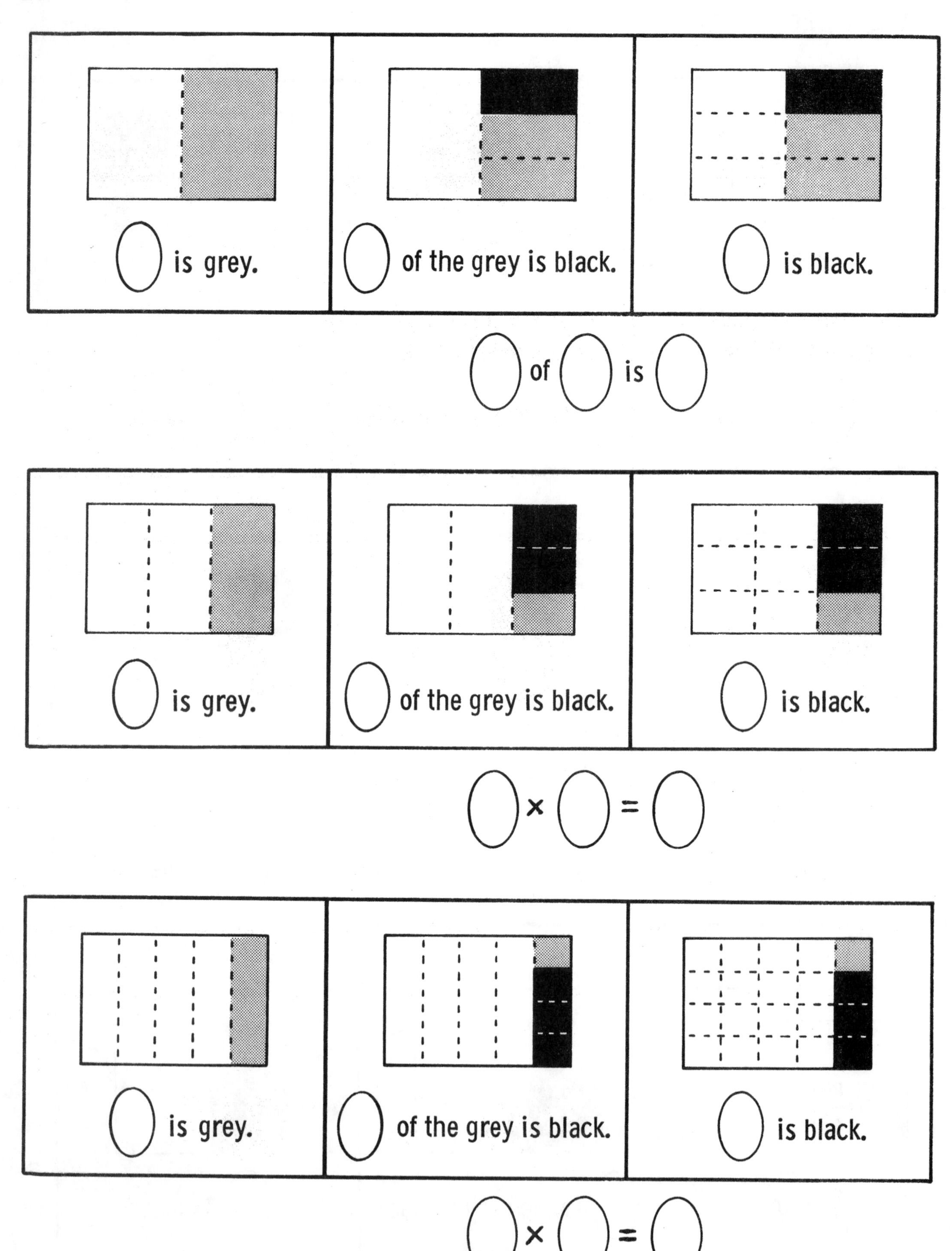
is grey.
of the grey is black.
is black.
of
is
is grey.
of the grey is black.
is black.
×
=
is grey.
of the grey is black.
is black.
×
=

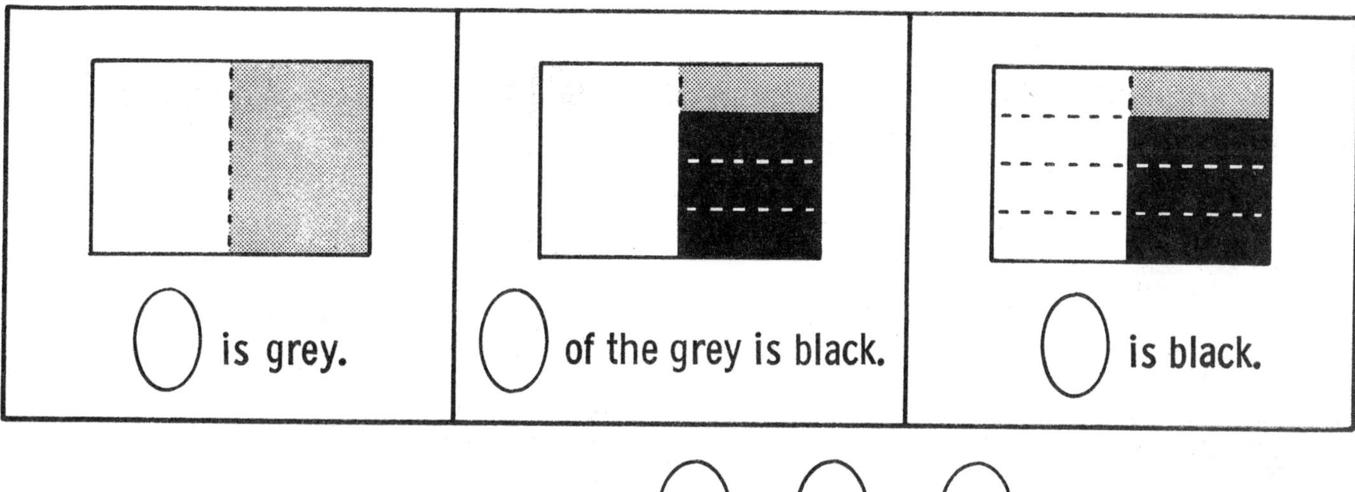

$\bigcirc \times \bigcirc = \bigcirc$

There is a way to multiply the fractions above without using pictures. Just multiply the numerators together and multiply the denominators together.

multiply numerators →
multiply denominators →

$$\frac{3}{4} \times \frac{1}{2} = \frac{3}{8}$$

Multiply the fractions below.

$\frac{4}{7} \times \frac{1}{3} =$ $\qquad$ $\frac{1}{3} \times \frac{1}{4} =$ $\qquad$ $\frac{1}{4} \times \frac{1}{3} =$

$\frac{1}{2} \times \frac{5}{12} =$ $\qquad$ $\frac{2}{9} \times \frac{2}{7} =$ $\qquad$ $\frac{2}{5} \times \frac{2}{5} =$

$\frac{2}{9} \times \frac{5}{9} =$ $\qquad$ $\frac{3}{7} \times \frac{1}{8} =$ $\qquad$ $\frac{1}{8} \times \frac{3}{7} =$

$\frac{5}{7} \times \frac{1}{2} \times \frac{1}{3} =$ $\qquad$ $\frac{7}{9} \times \frac{5}{8} =$ $\qquad$ $\frac{5}{8} \times \frac{7}{9} =$

$\frac{2}{5} \times \frac{3}{11} \times \frac{2}{5} =$ $\qquad$ $\frac{1}{2} \times \frac{3}{5} \times \frac{3}{7} =$ $\qquad$ $\frac{3}{5} \times \frac{1}{2} \times \frac{1}{2} \times \frac{3}{4} =$

Multiplying and Simplifying

Multiply and then simplify. Always make sure that you have found the simplest form for the answer.

$\frac{1}{2} \times \frac{2}{5} = \mathbf{\frac{2}{10}} = \mathbf{\frac{1}{5}}$

$\frac{2}{5} \times \frac{3}{4} =$

$\frac{3}{4} \times \frac{2}{3} =$

$\frac{4}{7} \times \frac{1}{2} =$

$\frac{6}{7} \times \frac{1}{3} =$

$\frac{8}{11} \times \frac{1}{2} =$

$\frac{8}{9} \times \frac{3}{4} =$

$\frac{1}{7} \times \frac{7}{8} =$

$\frac{2}{3} \times \frac{3}{10} =$

$\frac{3}{8} \times \frac{1}{3} =$

$\frac{6}{7} \times \frac{1}{2} \times \frac{1}{2} =$

$\frac{2}{3} \times \frac{3}{4} \times \frac{1}{3} =$

Look carefully below. You can save yourself some work.

$\frac{3}{4} \times \frac{4}{5} =$

$\frac{4}{5} \times \frac{3}{4} =$

$\frac{2}{9} \times \frac{3}{10} =$

$\frac{3}{10} \times \frac{2}{9} =$

$\frac{7}{12} \times \frac{4}{5} =$

$\frac{4}{5} \times \frac{7}{12} =$

Word Problems

Below are some problems. To solve each problem you must multiply fractions. Follow these four steps:

1. Read the problem.
2. Write it as multiplication of two fractions.
3. Multiply and simplify.
4. State your answer in a sentence.

Larry ordered $\frac{3}{8}$ of a pizza. He gave Pat $\frac{1}{3}$ of his order. How much of a pizza did Pat get?

Problem: $\frac{1}{3} \times \frac{3}{8} = \frac{3}{24} = \frac{1}{8}$

Answer: Pat got $\frac{1}{8}$ of a pizza.

My friend finished the test in $\frac{3}{4}$ of an hour. I finished it in $\frac{2}{3}$ of that time. How long did it take me to finish the test?

Problem:

Answer:

Cookies

$\frac{2}{3}$ of the cookies are small.
$\frac{5}{7}$ of the small cookies are chocolate.
What fraction are small chocolate cookies?

Problem:

Answer:

$\frac{3}{5}$ of the books in the library are non-fiction. $\frac{1}{12}$ of those are biographies. What fraction of the books in the library are biographies?

Problem:

Answer:

Today it took Gloria three fourths of an hour to come home from school. One third of that time she rode the bus. What part of an hour was the bus ride?

Problem:

Answer:

In ten minutes Janet walked two thirds of a mile. Pedro walked five sixths as far as Janet in ten minutes. What fraction of a mile did Pedro walk in ten minutes?

Problem:

Answer:

November 18 - Math

$\frac{4}{5}$ of the class was present. $\frac{2}{3}$ of those present turned in homework papers. $\frac{1}{2}$ of the papers turned in were correct.

Questions:

A. What fraction of the class turned in homework?

B. What fraction of the class had correct homework?

A. Problem:

Answer:

B. Problem:

Answer:

Simplifying and Multiplying

You can save work in some multiplication problems if you simplify before you multiply.

Here's how to simplify before you multiply:

Step 1	Step 2	Step 3
Find one numerator and one denominator with a common factor.	Divide these numbers by their common factor.	Multiply.

3 divides into 3.
3 divides into 6.

$\frac{1}{\not{6}} \times \frac{\not{3}}{5} =$

3 into 3 is 1

$\frac{1}{{}_2\not{6}} \times \frac{\not{3}^{\,1}}{5} =$

3 into 6 is 2

$\frac{1}{{}_2\not{6}} \times \frac{\not{3}^{\,1}}{5} = \frac{1}{10}$

Simplify and multiply.

$\frac{{}^2\not{4}}{7} \times \frac{1}{\not{2}_{\,1}} = \mathbf{\frac{2}{7}}$ $\qquad \frac{1}{6} \times \frac{3}{4} =$ $\qquad \frac{2}{3} \times \frac{3}{5} =$

$\frac{7}{8} \times \frac{1}{7} =$ $\qquad \frac{5}{6} \times \frac{1}{15} =$ $\qquad \frac{8}{9} \times \frac{1}{2} =$

$\frac{5}{7} \times \frac{3}{10} =$ $\qquad \frac{3}{4} \times \frac{1}{3} =$ $\qquad \frac{4}{21} \times \frac{7}{9} =$

$\frac{2}{3} \times \frac{5}{6} =$ $\qquad \frac{1}{3} \times \frac{1}{3} \times \frac{12}{13} =$ $\qquad \frac{7}{10} \times \frac{5}{6} \times \frac{1}{2} =$

Here's another example:

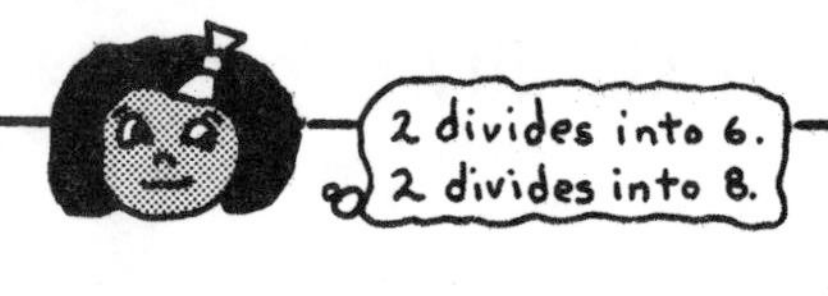

$$\frac{\not{6}}{7} \times \frac{1}{\not{8}} =$$

2 into 6 is 3.

$$\frac{{}^{3}\not{6}}{7} \times \frac{1}{\not{8}_{4}} =$$

2 into 8 is 4.

$$\frac{{}^{3}\not{6}}{7} \times \frac{1}{\not{8}_{4}} = \frac{3}{28}$$

Simplify and multiply

$\frac{9}{11} \times \frac{5}{6} =$ $\qquad$ $\frac{2}{5} \times \frac{1}{2} =$ $\qquad$ $\frac{6}{13} \times \frac{1}{4} =$

$\frac{8}{9} \times \frac{1}{6} =$ $\qquad$ $\frac{2}{7} \times \frac{3}{4} =$ $\qquad$ $\frac{1}{14} \times \frac{7}{8} =$

$\frac{5}{7} \times \frac{1}{10} =$ $\qquad$ $\frac{3}{10} \times \frac{5}{8} =$ $\qquad$ $\frac{11}{12} \times \frac{5}{11} =$

$\frac{1}{6} \times \frac{9}{10} =$ $\qquad$ $\frac{5}{6} \times \frac{7}{15} =$ $\qquad$ $\frac{1}{6} \times \frac{4}{5} =$

$\frac{5}{6} \times \frac{10}{11} =$ $\qquad$ $\frac{6}{7} \times \frac{1}{15} =$ $\qquad$ $\frac{9}{10} \times \frac{13}{15} =$

$\frac{2}{7} \times \frac{3}{10} =$ $\qquad$ $\frac{14}{15} \times \frac{1}{21} =$ $\qquad$ $\frac{3}{8} \times \frac{17}{33} =$

Sometimes you can divide two pairs of numbers to simplify a multiplication problem.

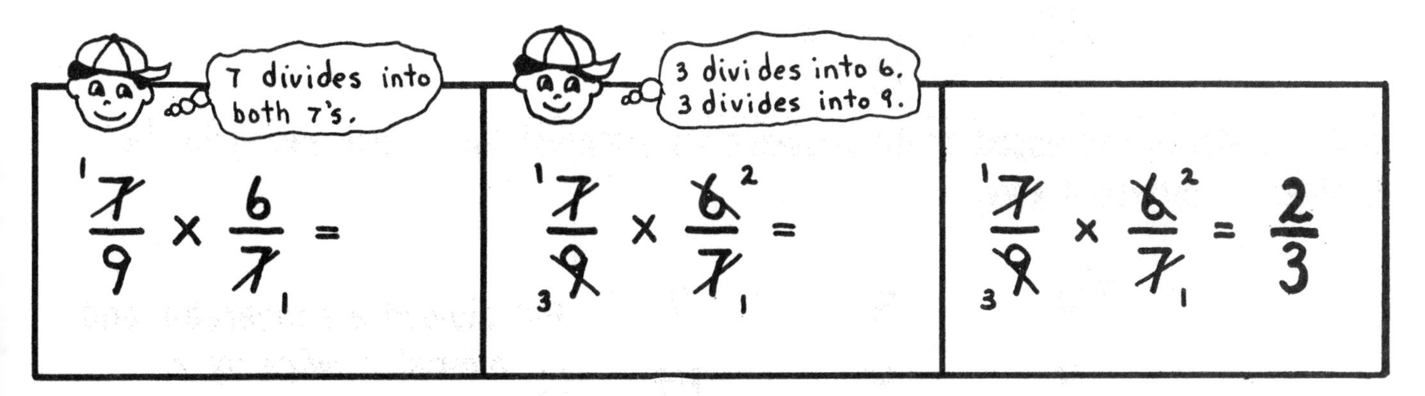

Divide two pairs of numbers and multiply.

$\frac{3}{15} \times \frac{5}{9} = \frac{1}{9}$ $\qquad \frac{3}{10} \times \frac{2}{3} =$ $\qquad \frac{5}{12} \times \frac{3}{5} =$

$\frac{3}{8} \times \frac{2}{9} =$ $\qquad \frac{2}{3} \times \frac{3}{4} =$ $\qquad \frac{6}{7} \times \frac{7}{15} =$

$\frac{2}{21} \times \frac{9}{10} =$ $\qquad \frac{5}{6} \times \frac{3}{25} =$ $\qquad \frac{5}{14} \times \frac{4}{5} =$

$\frac{3}{14} \times \frac{7}{9} =$ $\qquad \frac{1}{4} \times \frac{2}{3} \times \frac{9}{10} =$ $\qquad \frac{3}{5} \times \frac{10}{11} \times \frac{2}{9} =$

$\frac{5}{11} \times \frac{11}{5} =$ $\qquad \frac{7}{3} \times \frac{3}{7} =$ $\qquad \frac{3}{2} \times \frac{2}{3} =$

Pat and Sandy were asked to multiply:

$$\frac{4}{7} \times \frac{3}{8} =$$

Both students expressed their answers in simplest form, but each did the problem a different way.

Pat $\frac{\not{4}^{\,2\,1}}{7} \times \frac{3}{\not{8}_{\,4\,2}} = \frac{3}{14}$ Pat divided a numerator and denominator twice by 2.

Sandy $\frac{\not{4}^{\,1}}{7} \times \frac{3}{\not{8}_{\,2}} = \frac{3}{14}$ Sandy divided once by 4. 4 is the greatest common factor of 4 and 8.

Both answers are correct. Sandy's way is quicker.

Multiply and express your answer in simplest form. Try to divide using the greatest common factor. If you don't pick the greatest common factor, you will have to divide more than once.

$\frac{11}{30} \times \frac{15}{17} =$ $\frac{3}{8} \times \frac{4}{11} =$ $\frac{6}{7} \times \frac{5}{12} =$

$\frac{12}{17} \times \frac{1}{12} =$ $\frac{1}{12} \times \frac{8}{9} =$ $\frac{9}{13} \times \frac{5}{18} =$

$\frac{4}{9} \times \frac{9}{13} =$ $\frac{2}{27} \times \frac{9}{11} =$ $\frac{12}{13} \times \frac{5}{24} =$

Multiply. Express your answer in simplest form.

What is the product of $\frac{2}{7}$ and $\frac{14}{15}$? $\frac{2}{\not{7}_1} \times \frac{\not{14}^2}{15} = \frac{4}{15}$	Multiply $\frac{1}{3}$ and $\frac{12}{13}$.
What is $\frac{3}{7}$ of $\frac{4}{7}$?	What is $\frac{5}{6}$ times $\frac{3}{10}$?
Multiply $\frac{1}{6}$ by $\frac{3}{5}$.	Find the product of $\frac{2}{5}$, $\frac{3}{5}$, and $\frac{4}{5}$.

To multiply and express your answer in simplest form, you can multiply first and then simplify, or you can simplify first and then multiply. Although it is usually easier to simplify first and then multiply, you should get the same answer either way you do the problem.

Multiply first. Then simplify.	Simplify first. Then multiply.	Same answer?
$\frac{3}{7} \times \frac{7}{10} = \frac{21}{70} = \frac{3}{10}$	$\frac{3}{\not{7}_1} \times \frac{\not{7}^1}{10} = \frac{3}{10}$	(yes) no
$\frac{5}{9} \times \frac{4}{5} =$	$\frac{5}{9} \times \frac{4}{5} =$	yes no
$\frac{2}{7} \times \frac{5}{6} =$	$\frac{2}{7} \times \frac{5}{6} =$	yes no
$\frac{9}{10} \times \frac{2}{3} =$	$\frac{9}{10} \times \frac{2}{3} =$	yes no
$\frac{1}{2} \times \frac{2}{9} \times \frac{3}{5} =$	$\frac{1}{2} \times \frac{2}{9} \times \frac{3}{5} =$	yes no

Reciprocals

Multiply and then simplify.

$\frac{4}{5} \times \frac{3}{4} = \frac{12}{20} = \frac{3}{5}$

$\frac{2}{3} \times \frac{3}{4} =$

$\frac{2}{3} \times \frac{3}{2} =$

$\frac{2}{3} \times \frac{6}{7} =$

$\frac{4}{5} \times \frac{5}{4} =$

Simplify and then multiply.

$\frac{\overset{1}{\cancel{3}}}{5} \times \frac{2}{\underset{1}{\cancel{3}}} = \frac{2}{5}$

$\frac{3}{5} \times \frac{5}{3} =$

$\frac{3}{5} \times \frac{1}{3} =$

$\frac{2}{7} \times \frac{7}{2} =$

$\frac{5}{6} \times \frac{6}{5} =$

Draw a loop around each pair of fractions above whose product is 1.

Two fractions whose product equals one are called reciprocals.
$\frac{4}{3}$ and $\frac{3}{4}$ are reciprocals because $\frac{4}{3} \times \frac{3}{4} = 1$.
$\frac{4}{3}$ is the reciprocal of $\frac{3}{4}$. $\frac{3}{4}$ is the reciprocal of $\frac{4}{3}$.

Find the reciprocal needed to make 1.

$\frac{3}{4} \times \left(\frac{4}{3}\right) = 1$ $\qquad \frac{5}{2} \times \bigcirc = 1$ $\qquad \frac{1}{9} \times \bigcirc = 1$

$\bigcirc \times \frac{5}{6} = 1$ $\qquad \bigcirc \times \frac{3}{10} = 1$ $\qquad \bigcirc \times \frac{1}{4} = 1$

Find the reciprocal.

tricky!

$\frac{2}{7}$, $\bigcirc$ $\qquad \frac{9}{10}$, $\bigcirc$ $\qquad \frac{365}{487}$, $\bigcirc$ $\qquad \frac{1}{5}$, $\bigcirc$ $\qquad 5$, $\bigcirc$

Dividing Fractions

Reciprocals are used to divide by a fraction. Dividing by a fraction gives the same answer as multiplying by the fraction's reciprocal. Rewrite division by a fraction as multiplication by its reciprocal.

Follow the steps below to rewrite $\frac{5}{8} \div \frac{2}{3} =$ as multiplication.

Step 1 $\frac{5}{8} \div \frac{2}{3} = \frac{5}{8}$ Copy the first fraction.

Step 2 $\frac{5}{8} \div \frac{2}{3} = \frac{5}{8} \times$ Change division to multiplication.

Step 3 $\frac{5}{8} \div \frac{2}{3} = \frac{5}{8} \times \frac{3}{2}$ Change the second fraction to its reciprocal.

Fill in the missing numerators and denominators below in order to rewrite each division problem as multiplication by the reciprocal.

$\frac{4}{5} \div \frac{3}{7} = \frac{4}{5} \times \frac{7}{\quad}$ $\frac{1}{6} \div \frac{2}{3} = \frac{1}{6} \times \frac{3}{\quad}$

$\frac{2}{5} \div \frac{3}{4} = \frac{2}{5} \times \frac{\quad}{3}$ $\frac{3}{8} \div \frac{2}{5} = \frac{3}{\quad} \times \frac{5}{\quad}$

$\frac{5}{6} \div \frac{1}{2} = \frac{\quad}{6} \times \frac{2}{\quad}$ $\frac{1}{7} \div \frac{3}{4} = \frac{1}{7} \times \frac{\quad}{\quad}$

Circle the correct way to rewrite each division problem. Follow the steps above. Only one of the four choices is correct.

$\frac{1}{4} \div \frac{2}{3} =$ $\frac{4}{1} \times \frac{2}{3}$ $\frac{1}{4} \times \frac{2}{3}$ $\frac{1}{4} \times \frac{3}{2}$ $\frac{4}{1} \times \frac{3}{2}$

$\frac{2}{5} \div \frac{3}{7} =$ $\frac{2}{5} \times \frac{7}{3}$ $\frac{2}{5} \times \frac{3}{7}$ $\frac{5}{2} \times \frac{7}{3}$ $\frac{5}{2} \times \frac{3}{7}$

$\frac{1}{2} \div \frac{3}{5} =$ $\frac{2}{1} \times \frac{5}{3}$ $\frac{1}{2} \times \frac{3}{5}$ $\frac{2}{1} \times \frac{3}{5}$ $\frac{1}{2} \times \frac{5}{3}$

To divide by a fraction, rewrite the problem as multiplication by the fraction's reciprocal. Then multiply. Here are the steps:

1. Copy the first fraction.
2. Change division to multiplication.
3. Change the second fraction to its reciprocal.
4. Multiply.

Don't simplify until you have rewritten the problem as multiplication.

$\frac{1}{4} \div \frac{2}{3} = \frac{1}{4} \times \frac{3}{2} = \frac{3}{8}$

$\frac{1}{4} \div \frac{5}{7} =$

$\frac{2}{5} \div \frac{3}{7} =$

$\frac{2}{3} \div \frac{3}{4} =$

$\frac{3}{5} \div \frac{2}{3} =$

$\frac{1}{11} \div \frac{3}{4} =$

$\frac{4}{27} \div \frac{1}{2} =$

$\frac{1}{9} \div \frac{2}{13} =$

$\frac{2}{15} \div \frac{3}{11} =$

$\frac{5}{12} \div \frac{4}{7} =$

$\frac{1}{8} \div \frac{2}{3} =$

$\frac{4}{9} \div \frac{3}{5} =$

$\frac{1}{6} \div \frac{5}{7} =$

$\frac{5}{21} \div \frac{6}{11} =$

After you rewrite the division problems below, simplify and then multiply.

$\frac{2}{5} \div \frac{3}{5} = \frac{2}{\cancel{5}_1} \times \frac{\cancel{5}^1}{3} = \frac{2}{3}$ $\frac{3}{8} \div \frac{1}{2} =$

$\frac{5}{21} \div \frac{3}{7} =$ $\frac{2}{3} \div \frac{5}{6} =$

$\frac{3}{10} \div \frac{2}{5} =$ $\frac{6}{13} \div \frac{3}{4} =$

$\frac{5}{7} \div \frac{10}{11} =$ $\frac{3}{14} \div \frac{2}{7} =$

$\frac{7}{12} \div \frac{7}{8} =$ $\frac{8}{15} \div \frac{4}{5} =$

$\frac{3}{8} \div \frac{3}{4} =$ $\frac{4}{9} \div \frac{8}{15} =$

Divide. Express your answer in simplest form.

$\frac{1}{5} \div \frac{1}{4} =$ $\frac{1}{8} \div \frac{1}{4} =$

$\frac{2}{9} \div \frac{2}{3} =$ $\frac{7}{12} \div \frac{5}{6} =$

$\frac{2}{7} \div \frac{4}{5} =$ $\frac{1}{2} \div \frac{3}{4} =$

Multiplying and Dividing

Multiply or divide. Express your answer in simplest form.

$\frac{4}{7} \times \frac{5}{12} =$

$\frac{3}{8} \div \frac{4}{5} =$

$\frac{2}{3} \div \frac{4}{5} =$

$\frac{10}{11} \times \frac{2}{5} =$

$\frac{6}{11} \div \frac{2}{3} =$

$\frac{3}{8} \times \frac{4}{5} =$

$\frac{2}{9} \times \frac{6}{7} =$

$\frac{18}{19} \times \frac{5}{6} =$

$\frac{20}{33} \times \frac{11}{30} =$

$\frac{2}{7} \div \frac{6}{7} =$

$\frac{5}{6} \div \frac{8}{9} =$

$\frac{3}{4} \times \frac{7}{12} =$

Multiplication or division? Fill in × or ÷ to make each statement true.

$\frac{1}{7} \div \frac{2}{5} = \frac{5}{14}$

$\frac{7}{17} \quad \frac{2}{3} = \frac{14}{51}$

$\frac{1}{2} \quad \frac{1}{2} = 1$

$\frac{1}{7} \quad \frac{2}{5} = \frac{2}{35}$

$\frac{7}{17} \quad \frac{2}{3} = \frac{21}{34}$

$\frac{1}{2} \quad \frac{1}{2} = \frac{1}{4}$

$\frac{3}{5} \quad \frac{2}{3} = \frac{2}{5}$

$\frac{7}{12} \quad \frac{3}{4} = \frac{7}{16}$

$\frac{2}{13} \quad \frac{4}{5} = \frac{5}{26}$

Vocabulary Review

Make a true statement using one word or phrase from the box.

simplify - equal parts - reciprocal - whole numbers - denominator multiplying - factor - simplest form - GCF - fractions - numerators mixed numbers - equal fractions - greatest common factor

6 is a ________________ of 12.

$\frac{5}{6}$ is the ________________ of $\frac{6}{5}$.

$18\frac{3}{4}$, $5\frac{1}{2}$, $3\frac{1}{3}$ are examples of ________________.

$\frac{20}{36}$, $\frac{10}{11}$, $\frac{4}{5}$ all have even ________________.

I can ________________ $\frac{8}{12}$ to $\frac{2}{3}$.

The ________________ of 12 and 18 is 6.

Fractions can be shown by dividing a unit into ________________.

12 is the ________________ of 12 and 24.

$\frac{3}{5}$, $\frac{18}{5}$, $\frac{5}{5}$ all have the same ________________.

Dividing a fraction by a fraction is the same as ________________ the fraction by the reciprocal of the second fraction.

306, 2, 5 are all ________________.

$\frac{4}{9}$, $\frac{10}{3}$, $\frac{7}{8}$ are all ________________.

$\frac{4}{10}$, $\frac{8}{20}$, $\frac{2}{5}$ are ________________; only $\frac{2}{5}$ is in ________________.

Practice Test - Key To Fractions Book 2

Name ______________________

Date ______________________

Make equal fractions.

$\frac{5}{7} = \frac{10}{\quad}$ $\frac{3}{4} = \frac{\quad}{16}$

Circle the fraction in simplest form.

$\frac{7}{14} = \frac{3}{6} = \frac{1}{2} = \frac{9}{18} = \frac{4}{8}$

numbers	factors	common factors	GCF
12	___, ___, ___, ___, ___, ___	___, ___, ___	___
20	___, ___, ___, ___, ___, ___		

Find the GCF.

$\frac{7}{21}$ GCF is ___.

$\frac{15}{21}$ GCF is ___.

Simplify.

$\frac{6 \div 2}{16 \div 2} =$ $\frac{4}{12} =$ $\frac{12}{18} =$ $\frac{10}{10} =$

$\frac{15}{30} =$ $\frac{15}{20} =$

Extra Credit:

$\frac{100}{250} =$

GCF stands for the ______________ ______________ ______________.

Multiply.

$\frac{3}{5} \times \frac{1}{4} =$ $\frac{5}{7} \times \frac{3}{4} =$ $\frac{1}{2} \times \frac{1}{2} =$

Multiply and then simplify.

$\frac{8}{9} \times \frac{1}{2} =$ $\frac{3}{4} \times \frac{2}{3} =$ $\frac{7}{10} \times \frac{5}{6} =$

Practice Test - Page 2

One snowy day, $\frac{3}{4}$ of Ms. Stapler's class was late to school. $\frac{1}{3}$ of the late students brought notes. What fraction of Ms. Stapler's class brought late notes that day?

Problem:

Answer:

Simplify and then multiply.

$\frac{2}{3} \times \frac{3}{7} =$ $\frac{1}{8} \times \frac{2}{5} =$ $\frac{4}{5} \times \frac{5}{8} =$

Multiply. Express your answer in simplest form.

$\frac{5}{7} \times \frac{1}{5} =$ $\frac{3}{4} \times \frac{5}{8}$ $\frac{2}{5} \times \frac{1}{2} \times \frac{3}{4} =$

Find the reciprocal.

$\frac{3}{4}$, ◯ $\frac{8}{9}$, ◯ $\frac{37}{100}$, ◯ $\frac{7}{10} \times$ ◯ $= 1$

Divide. Remember to rewrite each problem as multiplication by the reciprocal.

$\frac{3}{8} \div \frac{2}{5} =$ $\frac{1}{7} \div \frac{1}{2} =$

Divide. Express your answer in simplest form.

$\frac{3}{8} \div \frac{1}{2} =$ $\frac{1}{7} \div \frac{1}{7} =$

Key to Fractions® workbooks

Book 1: Fraction Concepts
Book 2: Multiplying and Dividing
Book 3: Adding and Subtracting
Book 4: Mixed Numbers

Answers and Notes for Books 1–4
Reproducible Tests for Books 1–4

Also available in the Key to...® series

Key to Decimals®
Key to Percents®
Key to Algebra®
Key to Geometry®
Key to Measurement®
Key to Metric Measurement®
The Key to Tracker®, the online companion for the
Key to Fractions, Decimals, Percents, and Algebra workbooks

culum Press
THEMATICS EDUCATION

PRINTED WITH SOY INK

ISBN 978-0-913684-92-4